ENERGY

Rebecca Woodbury, Ph.D., M.Ed.

I0059381

Gravitas Publications Inc.

ENERGY

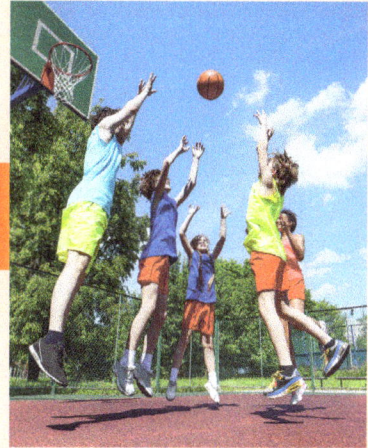

Illustrations: Janet Moneymaker

Copyright © 2024 by Rebecca Woodbury, Ph.D., M.Ed.

All rights reserved. No part of this publication may be reproduced, stored in a retrieval system, or transmitted, in any form or by any means, electronic, mechanical, photocopying, recording, or otherwise, without prior written permission from the publisher. No part of this book may be reproduced in any manner whatsoever without written permission.

Energy
ISBN 978-1-950415-21-2

Published by Gravitas Publications Inc.
Imprint: Real Science-4-Kids
www.gravitaspublications.com
www.realscience4kids.com

RS4K

Photo credits: Cover & title page: By Denis, AdobeStock; Above, By Sergey Novikov, AdobeStock; Page 3, Photo by Vitaly Gariev on Unsplash; Page 17, By Monkey Business, AdobeStock; Page 18: Cheese, Photo by Andra C Taylor Jr on Unsplash; Egg, Image by Pexels from Pixabay; Bread, Image by Graciela Moreno from Pixabay; Page 19, By Sergey Novikov, AdobeStock; Page 21, 1. By mamahoohooba, AdobeStock; 2. Photo by Loija Nguyen on Unsplash; 3. Image by Jasmine Lin from Pixabay; 4. Photo by J D on Unsplash; 5. Image by Aline Ponce from Pixabay; 6. By intararit, AdobeStock; 7. By warloka79, AdobeStock; 8. By New Africa, AdobeStock; 9. By SeventyFour, AdobeStock

You may have heard adults

say things like,

"I am so tired and out of energy."

Or...

- 4 -

Or...

Or...

But what is **energy**?

In **physics**, **energy** is

what is needed to do **work.**

Review: WORK

Work happens when a
force moves an object.

He is doing a lot of work!

Review: FORCE

Force is any action that changes:

- The **location** of an object,

- The **shape** of an object,

- **How fast or how slowly** an object is moving. (This is called the **speed** of an object.)

Putting it all together...

Energy is needed to do **work**.

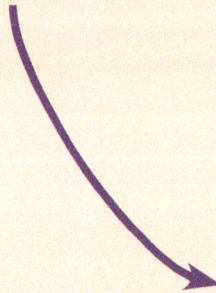

Work happens when a **force** moves an object.

Force is any action that changes:

- The **location** of an object,

- The **shape** of an object,

- **How fast or how slowly** an object is moving. (This is called the **speed** of an object.)

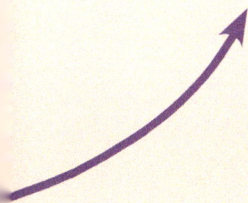

Take a look!

Remember that pulling a wagon up a hill is doing work.

But how do you pull

a wagon up a hill?

But how do your muscles get the **energy** to pull a wagon?

From the food you eat!

Energy comes from the cheese,
bread, or eggs you eat.

With energy you can do work.

They are doing work!

Energy comes in many forms.

- Food gives energy to our bodies.

- Batteries give energy to toys.

- Gasoline gives energy to cars.

Cheese gives energy to my body!

1

6

2

4

3

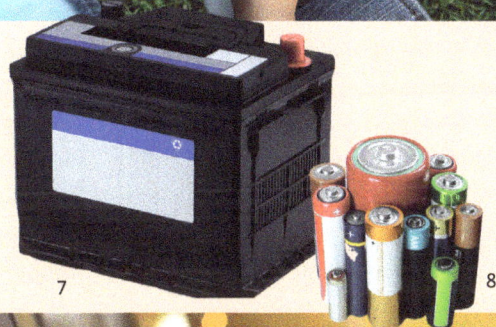

7

8

5

9

How to say science words

crisis (CRIY-suhs)

energy (E-nuhr-jee)

force (FAWRS)

location (loh-KAY-shun)

muscle (MUH-suhl)

physics (FIZ-iks)

shape (SHAYP)

speed (SPEED)

work (WERK)

www.ingramcontent.com/pod-product-compliance
Lightning Source LLC
Chambersburg PA
CBHW040149200326
41520CB00028B/7547